Real Science-4-Kids

Chemistry Connects to
History
Workbook Level I A

Rebecca W. Keller, Ph.D.

Cover design: David Keller
Opening page: David Keller
Illustrations: Janet Moneymaker, Rebecca Keller
    (unless otherwise noted)
Support Writer: Don Calbreath
Editing: Angie Sauberan
Page layout: Kimberly Keller

Copyright © 2008 Gravitas Publications, Inc.

Real Science-4-Kids/ Kogs-4-Kids™ : Chemistry Connects to History: Level I A

ISBN: 9780979945960

Published by Gravitas Publications, Inc.
P.O. Box 4790
Albuquerque, NM
87196-4790
www.gravitaspublications.com

Printed in the United States of America

Special thanks to G.E. McEwan for valuable input.

Gravitas
Publications Inc.

I Introduction
History

I.1 The history of science

Who were the first "scientists?" Where did they live? What did they discover? How did they make discoveries? All of these are questions about the history of science. Science didn't just happen overnight, but was developed over a long period of time. Science has a *history*.

The word "history" comes from the Greek word *historia,* which means "a learning by inquiry." History is the study of events that happened in the past, and it is the building of a narrative, or story, about those events. Historians look for documents and artifacts that tell them what happened a long time ago. Then, using the documents and artifacts that they have found, they try to piece together a story. Science has been developing and changing for many centuries, and as a result, science has a rich and exciting history.

I.2 Historians use documents and artifacts

Historians that study the history of science use all kinds of information to help them understand how modern science was developed. For example, historians might look at the notebook that Isaac Newton used when he was trying to understand how light travels through a prism. Historians might look at writings from ancient philosophers. They might look at the languages that various peoples used to describe the natural world. They might look at drawings people made about inventions people used. They might look at drawings that show what ancient people thought about the stars. They might also look at what other historians wrote in the past so that they can get an idea about how an invention or idea was viewed during an earlier time period. By using all of these types of information, historians try to determine what happened in the past, and they try to determine how the things we know today came about. We will learn about some of the tools historians use to understand the past, and we will also look at the historical events that helped to shape modern science.

I.3 Historians interpret events

Historians do more than just look at the events of the past. They try to understand what the events **mean**. That is, they try to interpret the past in order to better understand how the past affects the present and how the past might impact the future. It is very important to have a solid understanding of how our history was shaped. Many times, people forget history, and when they run into the same problems that earlier people encountered, they often make the same mistakes. If we don't remember our history, we may repeat the same mistakes without ever learning from our past.

Sometimes historians disagree about how to interpret the past. Just like scientists disagree about how to interpret data, historians disagree about how to interpret events. Just like in science, in history, it is always better to read several different interpretations of an event, and not just one. It is helpful to view all of the various viewpoints, and after you have studied them all, you can then come to your own conclusions about what you think they mean.

I.4 What you will learn

In this workbook, you will take a look at both the people and the historical events that helped shape modern science. Science has a history, and because scientific discoveries are made by people, it is the people that make the history. Because scientists are different people with different ideas about how to interpret scientific information, and because scientific information changes over time as more facts become available, heated arguments can erupt as a result of new discoveries. As you will learn by exploring the history of science, science grows when new facts and ideas compete with old facts and ideas. The new facts and ideas challenge old theories. The challenges presented to old theories can sometimes drastically change the conclusions that scientists make.

Science is a dynamic endeavor; it is full of exciting arguments and controversies. You will learn by exploring the history of science that the arguments and controversies fuel new scientific discoveries.

I.5 Discussion questions

1. Every person has a history. Write a short time line of your "history," starting with when you were born and continuing up until today. Place at least four or five events on your time line. For example, your time line might include events like "My First Birthday," "My Aunt's Wedding When I Was Two," etc.

Born

Today

2. Think of some resources that you might use to reconstruct your history, and list them below. (For example, you might use a birth certificate, photos, etc.)

_____

_____

_____

_____

_____

_____

3. Think about the histories of your grandfather and your grandmother. What resources could you use to reconstruct their histories?

_____

_____

_____

_____

_____

_____

_____

4. Think about your great-great-great-grandmother and your
    great-great-great-grandfather. What resources could you use to
    reconstruct their histories?

    _____

    _____

    _____

    _____

    _____

    _____

5. Do you think it is easier to reconstruct your history or the
    histories of your great-great-great-grandparents? Why?

    _____

    _____

    _____

    _____

    _____

    _____

    _____

1

The Elements
History

1.1 Introduction

1.2 The alchemists

1.3 Alchemy meets experiment

1.4 Organizing the elements

1.5 Activities

1.1 Introduction

We don't know when humans first started thinking about chemistry. However, every civilization we know about had some knowledge of silver and gold; these two elements have always been used as money.

Hammurabi, the sixth king of Babylon (1792–1750 B.C.), made a list of the elements that people used to make things. These elements were gold, silver, mercury, lead, tin, iron, and copper. His astrologers matched the elements with the sun, moon, and planets.

The ancient Roman Empire used gold and silver for money. They used lead for their water pipes, cups, and dishes. Lead can cause brain damage over time, and many people believe that the strange behavior exhibited by some of the later Roman rulers was caused, in part, by lead poisoning. For centuries, humans used the different elements, but they did not try to figure out what they were. A Greek philosopher named **Democritus** (*circa* 460 – *circa* 370 B.C.) first proposed that matter exists in the form of extremely small particles that he called "atoms." But no one believed him! Everyone thought that **Aristotle** (384–322 B.C.) was right and that matter was made of air, water, fire, and earth. It took almost 2000 years before atoms were considered the basic unit of all matter.

1.2 The alchemists

So for 2000 years, everyone believed that all things were made of air, fire, water, and earth. What happened to change this belief? Why do we know today that all things are made of atoms?

From Aristotle's time forward, people still thought about what things were made of, and they "experimented." There wasn't really a scientific method to their experimenting. That came later. But they were able to learn a lot about the properties of matter by "playing" with it. Some of these early experimenters were the **alchemists**. Alchemists were not considered to be true chemists because they did not approach their work scientifically. But they did play with the properties of matter. They believed that they could turn some things, like lead, into other things, like gold. A lot of what they tried was based on magic and didn't work. In fact, they never got any lead to turn into gold. Often they would go to a king and ask for a lot of money, and in return, they would promise the king that they could make gold out of lead. Of course, this never happened. Very often, the king would get angry and put the alchemists in prison (or worse). Sometimes the alchemists would just leave town with the king's money.

Although the alchemists were never successful at turning lead into gold, they did learn quite a lot about the properties of matter. They found out which things would burn, which things had a particular taste, and which things would cause bubbles if mixed with other things. Through this process, they collected lots of information about the properties of various elements.

1.3 Alchemy meets experiment

The alchemists didn't think that everything was made of air, water, fire, and earth. They thought that everything was made of mercury, sulfur, and salt! So already, Aristotle's four basic substances were being challenged by the alchemists. But the alchemists weren't right either. By the late 16th and early 17th centuries, modern scientific thinking began to take shape. Philosophy and invention started coming together, and many "thinkers" began thinking about how to do good scientific experiments.

Sir Robert Boyle (1627-1691)

One such thinker was Sir Robert Boyle. He challenged both Aristotle's four substances and the alchemists' three substances. Robert Boyle did not know what the basic substances were, but he argued that both the alchemists and those who believed the Aristotelian view were wrong. Boyle performed experiments to prove his ideas, and he made many contributions to chemistry. He used elaborate glassware to test the properties of air and fire, and

he figured out the fundamental gas laws. He helped to show that different kinds of atoms could combine to form molecules.

Boyle believed in running experiments to see what would actually happen. One of his experiments produced oxygen, but he was unaware of what he had produced. He thought he had produced something else.

Joseph Priestley (1733-1804)

Priestley never took a science course. He enjoyed playing around with different things. After he met Benjamin Franklin, Priestly began to get more interested in science. He discovered carbon dioxide, and he invented the first soda pop. Carbon dioxide is what makes the fizz in soft drinks. Laughing gas (used for anesthesia) was another of his many discoveries. He also did experiments with oxygen.

A.L. Lavoisier (1743–1794)

Antoine Lavoisier was a French scientist who was a friend of Priestley's.

He did some work with oxygen after Priestly told him about it. Lavoisier tried to take credit for the discovery of oxygen.

Lavoisier showed that hydrogen and oxygen could be "burned" in order to make water. Lavoisier believed in the experimental method, as did Boyle and Priestley. He called laboratory work "the torch of observation and experiment." This "torch" shed light on scientific facts. Lavoisier wrote a famous book on chemistry, which organized a lot of useful information.

Lavoisier was very rich, and that fact made trouble for him during the French Revolution. The common people took him prisoner and had him executed. Lavoisier was one of the many scientists who earned the title "The Father of Chemistry."

John Dalton (1766-1844)

By the early 1800s, it was well established that air, fire, water, and earth were not the basic substances. This paved the road for the work by John Dalton.

Dalton was a British schoolteacher for most of his life. He first became interested in science by studying the weather. This research then led him to discover some of the gas laws. John Dalton revived the hypothesis for the atomic theory of elements that had been proposed by Democritus some 2000 years earlier. In his published work from 1808-1827, called

A New System of Chemical Philosophy, Dalton proposed that all elements were made of atoms. He also proposed that each element had its own atomic weight. The atomic weight, he said, is proportional to the size of the atom (the number of protons and neutrons) that makes up the element—which is exactly what we know today. Dalton drew the first table of elements. In the table, he described the arrangement of the atoms in several elements, and he provided their atomic weights. Dalton did not know all of the elements that we know today, but he laid the foundation for future study. His contributions to the field of chemistry were significant. John Dalton is known as the "Father of Modern Chemistry." Dalton's atomic theory tried to explain some basic properties of atoms. He had the right idea, but several points in his theory were later proven incomplete.

Amedeo Avogadro (1776-1856)

Avogadro started his adult life as a lawyer. He later became interested in science, and he began his scientific studies in the field of electricity. He studied how atoms combine to form molecules. Avogadro found that "atoms" of nitrogen and "atoms" of oxygen were actually molecules. Molecules of nitrogen contain two N atoms, and molecules of oxygen contain two O atoms. He also found that equal volumes of gases contain the same number of molecules.

1.4 Organizing the elements

If you enjoy going to the library, you probably like to read about different subjects. When you go to a library, you might want to find a special book. But it will be hard to find that special book if there is no system to the way the books are shelved. Librarians have ways to organize books so that we can go directly to the ones we want. Chemists also like to organize information so that they can quickly find what they need. In 1869, **Dmitri Mendeleev** (1834-1907) expanded Dalton's table, and scribbling into his notebook, developed the first version of our modern periodic table. Mendeleev was born in 1834 in Tobolsk, Siberia (Russia). He was a chemist, and he carried with him cards that had the names and weights of the 63 known elements written on them. He thought about the elements and their weights a great deal. After much thinking, he decided to arrange the elements into a chart, based on their atomic masses. He published his chart in a book called

Principles of Chemistry in 1869. He left spaces in his chart because he thought that some elements were missing, and he was right! With his table, he was able to predict a few of the elements that were missing, and while he was still living, the next three elements were indeed discovered. His table gave other scientists the information they needed to find the missing elements. Those missing elements were exactly what Mendeleev predicted! He was famous for the success of his predictions.

Our periodic table today is much larger than Mendeleev's table (about 119 elements). Some of the newer elements have been created in the laboratory.

1.5 Activities

1. Complete the time line that is provided for you on page 18. Fill in the dates, and record the names of the early scientists who helped to discover atoms and to develop the periodic table.

2. Look up the word "alchemist" in the dictionary, and write the definition.

   _____

   _____

   _____

   _____

   _____

   _____

3. Why is John Dalton called the "Father of Modern Chemistry?"

   _____

   _____

   _____

   _____

Time line
the discovery of atoms

1900 A.D.

450 B.C.

4. Imagine that you are on a strange new planet. It has elements, just like planet Earth, but these elements are different. You come across a set of cards. These cards list the properties of the elements.

Cut out the individual cards on the last page in this chapter.

Try to arrange them into a periodic chart that resembles the one for planet Earth.

There is one card missing. Can you guess the atomic weight and properties?

atomic weight: _____

properties: _____

_____

_____

You are the first person to discover this new planet, and so you can name the planet anything you choose. Name your planet, and write a short description of the elements it contains.

Name: _____

_____

_____

_____

_____

Symbol: Ch

Name: Chocolatium

Atomic weight : 4

Properties:

brown powder; reacts violently with Sg

Symbol: Sp

Name: Spinachium

Atomic weight : 10

Properties:

inert; green; bitter taste rare

Symbol: Sg

Name: Sugarium

Atomic weight : 1

Properties:

white powder; sweet taste; reacts violently with Ch; abundant

---

Symbol: Bs

Name: Brusselsproutium

Atomic weight : 5

Properties:

inert; green; gives off foul odor if heated; rare

Symbol: B

Name: Butterarium

Atomic weight : 2

Properties:

yellow solid at room temp; reacts with Ts

Symbol: Pb

Name: Peanutbutterarium

Atomic weight : 7

Properties:

brown solid at room temp; reacts with Ts

Symbol: Hn

Name: Honeyrarium

Atomic weight : 6

Properties:

thick liquid at room temp; reacts violently with Ch

Symbol: Ts

Name: Toastonium

Atomic weight : 3

Properties:

speckled brown and white; solid at room temp; reacts with B and Pb

---

Symbol: Bc

Name: Biscuitium

Atomic weight : 8

Properties:

fluffy white solid; reacts with large amounts of B

2 Molecules and Bonds
History

2.1 Introduction

John Dalton's atomic theory was the start of modern chemistry. But there was a lot we did not know about the atom. Dalton believed the atom could not be further broken down into smaller parts, and he was right!

Today we know that there are parts to each atom. Some of these parts (the electrons) are very important for holding different atoms together and for forming ions. Other parts (neutrons) help make isotopes.

2.2 Electrons and protons

Michael Faraday (1791-1867) was an English chemist and physicist who showed that some materials could be dissolved in water. He discovered that when these materials dissolved in water, the solutions would conduct electricity. He also found out that these materials would separate into their basic

components when an electric current was passed through the solution. Through this work, he discovered the basic laws of **electrolysis**.

A Swedish scientist named **Svante Arrhenius** (1859-1927) further studied these processes of electrolysis. He showed that table salt (NaCl) would separate into two different components, each with a different electrical charge. Arrhenius called these materials "ions." The word *ion* comes from a Greek word meaning "the ones who move."

So at that time, he knew that some molecules could form electrically charged particles when dissolved. He learned that some of these particles had positive charges and that some had negative charges. But what he didn't know was that electrons caused the charges.

Electrons

The electron was discovered in 1897 by **Sir Joseph John Thomson** (1856-1940), an English physicist. Thomson studied mathematics in college. He then used math to study magnetic fields. He also studied some of the properties of this new "stuff" called electricity.

Thomson used the **cathode ray tube** in his experiments. The cathode ray tube is a long glass tube with wires attached to it. Air can be sucked out of the tube with a vacuum pump. When Thomson attached the wires of the tube to a supply of electricity, he could see a path of light going from one end of the tube to the other. Thomson and others looked at how magnetic fields affected the path of light. If no magnet was present, the light path would be straight.

When a magnet was put next to the tube, the light path would bend. From these experiments, Thomson discovered the electron. His research showed that the electron is very, very small and that it has a negative charge.

Protons

Because an atom has a negative charge (that is due to its electrons), it makes sense that in order for an atom to be balanced (not charged), it needs some kind of positively charged particle. Many scientists did experiments to look for these positive particles. Most people believe that **Ernest Rutherford** (1871-1937) developed the idea of the proton. Ernest Rutherford was born in New Zealand, and he later went to England to study. Rutherford learned that the proton is a hydrogen atom that is missing its electron. ("Proton" comes from a Greek word meaning "first.")

2.3 Ionic bonding

We can now explain ionic bonding. In an ionic bond, positive ions lose one or more electrons and are attracted to negative ions that have gained one or more electrons. Opposites attract, and the ions stay close to one another.

Amedeo Avogadro (1776-1856) was probably the first person to distinguish between atoms and molecules. He talked about particles called molecules that were made up of atoms. His theory dealt with only even numbers of atoms, and he did not say anything about odd numbers of atoms.

Gilbert N. Lewis (1875-1946) was an American chemist who first came up with the idea of the covalent bond. This type of bond is formed when electrons from two atoms are shared. Neither atom "owns" the electrons. In 1927, the **valence bond theory** was proposed. This theory says that the shared electrons are in the outermost **shell** of each atom (kind of like the outer layers of an onion). Only the outside electrons can participate in bonding, and at least two electrons have to be shared in order to have a covalent bond.

2.4 The idea of electron sharing

Once scientists knew about electrons in atoms, they could then ask how atoms were held together. Many compounds can form ions in a water solution, but many others cannot. In fact, millions of compounds do not even dissolve in water.

The idea of atoms connecting to form molecules was thought about for centuries. **René Descartes** (1596-1650) was a French philosopher who believed that molecules were held together by little hooks and eyes. He believed that some atoms had hooks and that others had eyes where the hooks could connect.

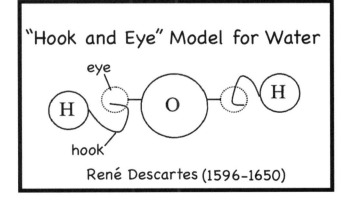

"Hook and Eye" Model for Water

René Descartes (1596-1650)

Many other scientists and philosophers thought about this problem. Many accepted some form of the "hook and eye" theory. John Dalton did analyses of different molecules and showed they could be made of collections of atoms.

Linus Pauling (1901-1994) was one of the most influential individuals to develop the idea of the covalent bond. In 1931, he wrote a paper called "The Nature of the Chemical Bond." This paper described his theories about covalent bonds. He later wrote a book called *The Nature of the Chemical Bond and the Structure of Molecules and Crystals.* In his paper and in his book, he described how molecules share electrons through covalent bonding.

Pauling won the Nobel Prize in Chemistry in 1954 for his chemical bond research. He also won the Nobel Peace Prize in 1962 for his efforts to stop some nuclear testing. He later developed the idea that very large doses of vitamin C can cure diseases. Although it is an interesting idea, vitamin C has not been found to cure diseases.

2.5 Models for molecules— chemistry and math

A lot of theory about chemical bonds is now done with math. These ideas are very complicated, but they can be very useful for many purposes. We won't go into these complicated ideas until you have taken calculus.

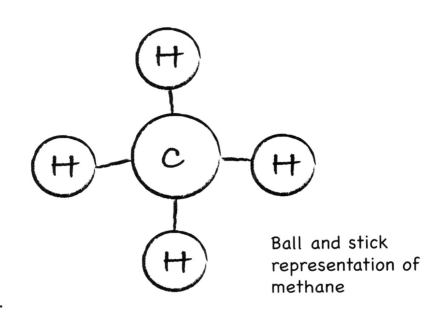

Ball and stick representation of methane

The picture of a molecule that we see on a page does not give the complete idea about that molecule. The picture is flat. We need to see a molecule in three dimensions in order to really understand how it will react.

August Wilhelm von Hofmann (1818-1892) was the first person to build models of molecules based on theory. He used sticks for chemical bonds and balls for the individual atoms. He did not know the real shapes, but he gave us a start in understanding the science of molecule shape.

Hermann Emil Fischer (1852-1919) was a German chemist, and he proposed some ideas about molecular shape for enzymes (biochemical molecules). He also developed a way of drawing (on paper) a two-dimensional representation of a three-dimensional molecule. In a Fischer projection, all of the bonds are drawn as vertical or horizontal lines. For methane (shown), the carbon atom is represented by the intersection of the horizontal and vertical lines. A Fischer projection is a representation of a three-dimensional molecule. The horizontal lines represent bonds that come forward, out of the page.

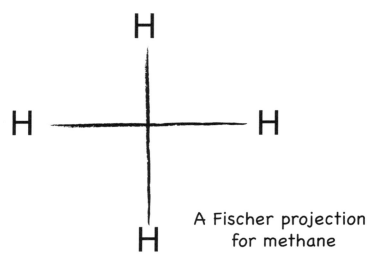

A Fischer projection for methane

2.5 Activity

On the last page of this chapter, there is a time line. Complete the time line from 1500 to 1970. On the time line, plot the dates of the achievements that were made by every scientist discussed in this chapter. If you want, you can look up the exact date of each achievement. Or you can estimate the dates by assuming that the achievements took place sometime in the middle of the scientists' lifetimes. Draw a "dot" to represent the time of each achievement, and write the scientists' names next to the dots.

When you finish the time line, answer the questions on the following page.

1. What is the time span between Descartes's philosophical ideas about how molecules bond to each other and the first experiments that began to uncover the nature of bonding?

_____

_____

_____

2. Explain what you observe on the time line regarding the dates of experimental discoveries. Are these dates clustered, or are they spread over many years?

_____

_____

_____

_____

3. Why do you think there is a long period of time between Descartes's philosophical ideas and the experimental discoveries?

_____

_____

_____

_____

4. Why do you think the experimental discoveries cluster within a relatively short period of time?

_____

_____

_____

_____

_____

Time line
the discovery of bonds

1970

1500

3 Chemical Reactions History

3.1 Introduction

The idea of chemical reactions arose in a time before history was recorded. Cooking uses a lot of chemical reactions. Some chemical reactions in cooking are good. They give us pizza. Other chemical reactions in cooking are not so good—like when we get burnt toast and have to start over again!

Ancient people who used animal skins to make clothes did chemistry. The ancient people treated the animal skins in order to make them

soft. During the treatment of the skins, chemical reactions occurred. The ancient people didn't know what the reactions were, but they knew that something worked. Ancient farmers who grew crops for food discovered other reactions. They knew

that there were materials that they could put in the ground, and they knew that those materials would help the plants grow. Today, we know what the chemicals are and how they work.

3.2 Reactions are used to study things

Scientists could not really study chemical reactions until they knew about atoms and molecules. Many of the early reactions that people explored helped them to learn what made up matter. They could do a reaction, and then observe what they got as products. That information told them what atoms were in the original compounds.

For example, **Henry Cavendish** (1731-1810) reacted different metals with hydrochloric acid (HCl). These displacement reactions resulted in the formation of a salt and in the release of hydrogen gas. He called the gas "inflammable air" because it burned.

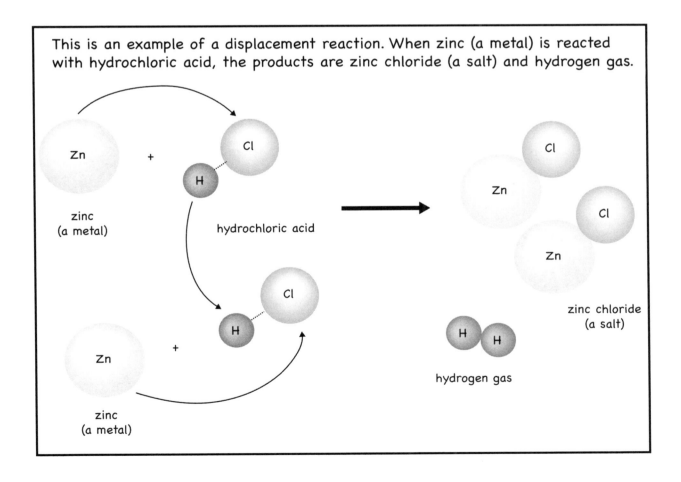

This is an example of a displacement reaction. When zinc (a metal) is reacted with hydrochloric acid, the products are zinc chloride (a salt) and hydrogen gas.

zinc (a metal) + hydrochloric acid + zinc (a metal) → zinc chloride (a salt) + hydrogen gas

Joseph Priestley (1733-1804) was another scientist who studied chemical reactions. Recall from chapter one that he never took any science courses. He first studied to be a preacher, but that didn't work out too well. So he decided to become a scientist. He went on to make many science discoveries.

One of Priestley's experiments involved heating a compound known as mercuric oxide. He did not know what it was made of. The experiment showed that oxygen was produced (and mercury, of course). He found that this gas helped things burn. Priestley studied reactions both in the test tube and in living systems. He studied the gas that plants produce. Priestly found that plants make the same gas that he had isolated by heating mercuric oxide.

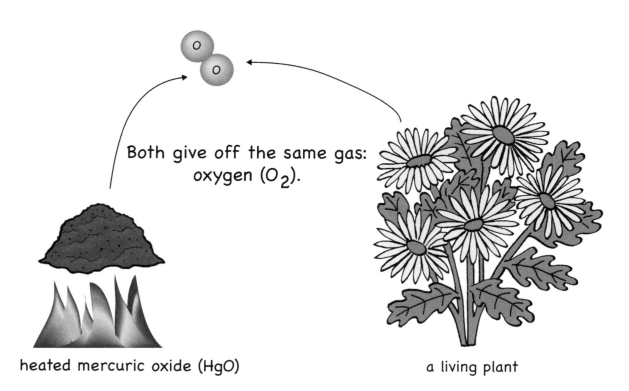

Both give off the same gas: oxygen (O_2).

heated mercuric oxide (HgO)   a living plant

Antoine Lavoisier (1743-1794) used this discovery to learn more about fire. Recall from chapter one that Lavoisier was the friend of Joseph Priestly. Like Priestly, Lavoisier was interested in learning more about gases. Lavoisier found that burning was a chemical reaction between the material being burned and oxygen. He also showed that hydrogen and oxygen could react together to form water. Lavoisier gave the name to the element hydrogen. Hydrogen comes from two Greek word roots that together mean "water producer." (In Greek, *hydro* means "water," and *gen* means "to produce.")

3.3 How to make your name famous

Many chemical reactions are called "name" reactions. These are famous reactions that everyone knows about. These reactions are very useful in a lot of chemical processes.

We use the name of the discoverer as a type of code. When we say "the Grignard reaction," that tells us what kind of chemical process is taking place. The Grignard reaction was discovered by the French chemist **François Auguste Victor Grignard** (1871- 1935). Not only does he have a chemical reaction named after him, but he also won the Nobel Prize in Chemistry for his discovery!

Other famous reactions are the "Cannizzaro reaction," named after the Italian chemist **Stanislao Cannizzaro** (1826–1910), and the "Claisen reaction," named after the famous German chemist **Rainer Ludwig Claisen** (1851–1930). Many of these reactions are specific to a branch of chemistry called **organic chemistry**. Organic chemistry is the chemistry of carbon compounds. This branch of chemistry has over 600 "name" reactions. Biochemistry (the chemistry of living systems) also has many "name" reactions. We will see more about some of these reactions in later chapters.

3.4 Making money from a "not-so-useful" reaction

Sometimes things just don't work out like you had planned. What you had expected to happen doesn't happen the way you had wanted. Chemists often have the same problem when they run experiments.

In 1968, a company was doing research on glues. They wanted to find chemical reactions that would make strong glues that would hold things together tightly. Some of the reactions made glues that just weren't very good. But scientists keep good records of all of their experiments. Even experiments that don't seem to work can tell you something useful. So, the reactions that produced weak glues were all written down in the research records.

Several years later, a friend of the guy who was making the glue needed notes that would stick in his book - but not get glued down. So together, these two friends got the idea for the "Post-it® Notes." They needed a weak glue that would allow the note to be pulled off easily. The scientists looked at their research notes and found a "failed" experiment that gave them just the glue they needed. In 1980, Post-its® became a product, and because of that "failed experiment," we can all write little "sticky notes" and put them on everything.

3.5 Activity

On the last page of this chapter, there is a time line. Complete the time line from 1700 to 1980. On the time line, plot the dates of the discoveries that were made by every scientist discussed in this chapter. If you want, you can look up the exact date of each discovery. Or you can estimate the dates by assuming that the discoveries took place sometime in the middle of the scientists' lifetimes. Draw a "dot" to represent the time of each discovery, and write the scientists' names next to the dots.

When you finish the time line, answer the questions on the following page.

1. Explain what you observe on the time line regarding the dates of experimental discoveries. Are these dates clustered, or are they spread over many years?

_____

_____

_____

_____

_____

2. Compare this time line to the time line in chapter two. What do you notice?

_____

_____

_____

_____

_____

3. Compare this time line to the time line in chapter one. Are there scientists who worked on more than one important issue?

_____

_____

_____

_____

4. Notice the time span between the discovery of "weak glue" and the creation of a new consumer product. What does this tell you about how scientific discoveries are turned into consumer products?

_____

_____

_____

_____

_____

_____

_____

5. What do you think would have happened if the "failed" weak glue experiment had been thrown away? What does this tell you about "failed" experiments?

_____

_____

_____

_____

_____

_____

_____

Time line
the discovery of chemical reactions

1980

1700

4 Acids and Bases

History

4.1 Introduction

Acids and bases have been used for many centuries. Even before people knew what they were, these types of compounds had many applications. Some of these compounds were used to get people clean. Some of these compounds were useful to the alchemists for experimentation.

Ancient civilizations used a base to make **soap**. The base they used is called lye. The lye came from the ashes that were leftover after wood was burned. Different kinds of oils or animal fats were heated with the lye, and this produced the soap that they used. This soap could get people clean, but it often damaged their skin.

Tablets from the Babylonian culture show that people knew how to make soap 2800 years ago. Some of the people used it to style their hair. Ancient German tribes also used soap for this purpose.

Some acids were known many centuries before today. An Iranian alchemist named **Abu Musa Jabir ibn Hayyan** (*circa* 721 - *circa* 815 A.D.) discovered hydrochloric acid by mixing salt (sodium chloride) with sulfuric acid. Jabir ibn Hayyan also developed **aqua regia** (royal water) by mixing nitric acid and hydrochloric acid. This material could dissolve gold easily. Thus, aqua regia was often used to determine whether or not a "gold" was real.

4.2 Modern ideas

In 1834, **Michael Faraday** (1791-1867) discovered that acids and bases are electrolytes. These compounds form ions when they are dissolved in water and can conduct electricity.

A Swedish chemist named **Svante Arrhenius** (1859-1927) took the next step in 1884. He believed that acids produced hydrogen ions in water and that bases made hydroxide ions when dissolved in water.

This was a useful theory for many years. But there was a problem with it. The theory only worked for hydrogen and hydroxide ions.

A better theory was developed in 1923. A Danish chemist, **Johannes Bronsted** (1879-1947), and a British chemist, **Thomas Lowry** (1874–1936), both came up with the same idea. They said that acids contributed protons and that bases accepted protons. This new theory made it possible to understand more chemical reactions. Eventually, other solutions and solvents were studied in order to better understand chemical bonding.

4.3 Distinguishing acids from bases

The first method employed to determine whether something was an acid or a base was **litmus paper**. The word *litmus* comes from an old Norse word meaning "to dye or color." Lichens provides the dye used in litmus paper. Litmus paper was first used by the Spanish alchemist **Arnaldus de Villa Nova** (*circa* 1235-1313) to test pH.

The pH Scale

The number of hydrogen ions in a solution can be very large or very small. After a while, the number can become very hard to follow. The pH scale was developed to make it easy to talk about hydrogen ions in a solution. Large amounts of hydrogen ions give us a low pH. Small amounts of hydrogen ions produce a high pH.

Sören Peter Lauritz Sörensen (1868-1939), a Danish chemist, developed the first pH scale in 1909. This scale made it easier to describe how many hydrogen ions were in a solution. He worked for a brewery in Sweden. Knowing the pH of the mixture enabled the brewery to produce a better product.

Indicator dyes

Many dyes have been extracted from a variety of plant and animal sources, and they have been used to color cloth and to make paint. Some of these compounds have also been used as dyes to test pH.

Scientists have found that different dyes, or *indicators*, give different colors when the pH is changed. We can often use these color changes to determine the pH of a solution. There are over thirty-five common indicators known and used today.

The pH meter

Measuring hydrogen ion concentration was complicated, and a quick way of finding pH was needed. There were early attempts to make a pH meter. In 1906, two scientists tried to make glass probes to measure pH directly. The glass had to be very thin, and therefore, it broke easily.

The first successful pH meter was built by **Arnold Beckman** (1900-2004) in 1934. He was a chemistry professor at the California Institute of Technology. He was asked to find a way to measure the acidity of lemon juice.

The California Fruit Growers Exchange grew most of the citrus fruit in California at that time. They needed a quick and easy way to see how acidic the fruit was. This information helped them figure out when to harvest the fruit.

After Beckman built the first pH meter, he went into business. The first meters went on the market in 1935. Many people believed there was a need for only about 600 pH meters in the world. Beckman proved them all wrong.

Over the next couple of decades, Beckman's company grew. Other science instruments were developed and sold. Beckman became a millionaire, and he was very generous with his money. During his lifetime, he contributed over $400 million dollars to science research and education.

4.4 Activity

On the last page of this chapter, there is a time line. On the time line, plot the dates of the discoveries that were made by every scientist discussed in this chapter. If you want, you can look up the exact date of each discovery. Or you can estimate the dates by assuming that the discoveries took place sometime in the middle of the scientists' lifetimes. Draw a "dot" to represent the time of each discovery, and write the scientists' names next to the dots.

After you complete the time line, please answer the questions on the following page.

1. Describe what you notice about the time span between the discovery of the first acid in 800 and the development of the modern pH meter in 1934.

_____

_____

_____

_____

2. What other discoveries were needed before the pH meter could be invented by Arnold Beckman?

_____

_____

_____

_____

_____

_____

_____

Time line
acids and bases

1950

800

5 Analysis
History

5.1 Introduction

5.2 Acid-base indicators and titration

5.3 The buret

5.4 The pH meter brings more accuracy

5.5 Where do graphs come from?

5.6 Activity

5.1 Introduction

What is **analysis**? In chapter five of Chemistry Level I, you learned how to **analyze** an acid-base reaction. You followed a method called **titration** to determine how many teaspoons of ammonia you would need to neutralize a given amount of vinegar. The method of titration is a type of analysis.

Learning how to analyze the world around us is something that scientists spend quite a lot of time doing. In fact, analyzing the world is one of the main tasks of science. Scientists look at the world around them and at everything it contains in order to determine the quantity, speed, height, weight, and length of the things in the world. Some of the very first analyses early scientists performed were on the earth itself. How big is it? How far is it from the sun? How

far is it from the moon? What is it made of? Using different tools and techniques, scientists do many different kinds of analyses. Many of these techniques are modern, but some have been developed over the course of centuries.

5.2 Acid-base indicators and titration

One of the ways you have learned to analyze an acid-base reaction is by using an acid-base indicator. Red cabbage juice is an easy acid-base indicator to use. There is a pigment in red cabbage called **anthocyanin**. This is a molecule that will turn different colors in different pH environments. There are other fruits and vegetables that contain anthocyanins, including eggplant, cherries, and grapes. Although no one knows for sure when red cabbage was first used as an acid-base indicator, the use of red cabbage as an indicator was first described in England around 1570.

Another common indicator for acid-base titrations is **phenolphthalein**. This indicator was discovered in 1871 by **Adolf von Baeyer** (1835–1917). He won the Nobel Prize in Chemistry in 1905 for his work with dyes. Phenolphthalein will turn from colorless to pink when acid is neutralized. Phenolphthalein was also used for many years as a laxative. It can also be used to test for the presence of blood. The dye was even used to change hair color in some Barbie dolls.

5.3 The buret

We need special equipment to measure liquids accurately. A piece of equipment used in titrations is called a **buret**. It is sometimes spelled "burette." The word comes from the French word for "vase." **Joseph Louis Gay-Lussac** (1778–1850) was a French chemist and physicist, and he is believed to have invented the first buret. The buret is a long glass cylinder with markings on the side. The markings are used to measure the volume of liquid in the buret. There is a stopcock that is used to let liquid out of the buret and into the solution being titrated.

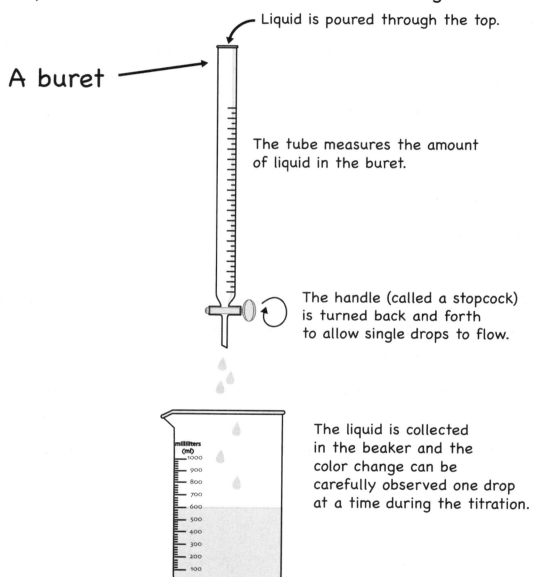

Liquid is poured through the top.

A buret

The tube measures the amount of liquid in the buret.

The handle (called a stopcock) is turned back and forth to allow single drops to flow.

The liquid is collected in the beaker and the color change can be carefully observed one drop at a time during the titration.

The first titrations used phenolphthalein to determine when the acid was neutralized. A base was added to the acid until the phenolphthalein turned a very light pink. The color change indicated that the acid had been neutralized.

5.4 The pH meter brings more accuracy

Titration using an indicator is useful, but it can be inaccurate. The degree of pink that people see is subjective. That is, everyone may see "pink" differently, and when this happens, extra base can be added by accident.

There are people who cannot see the pink color change. These people are color-blind, and they cannot detect subtle differences in color change. Many labs are required to test employees for color-blindness before they can run certain tests.

Use of a pH meter helps with these problems. We saw in chapter four that **Arnold Beckman** invented the first pH meter. Because the pH meter can read the pH value directly, no one has to decide how pink the solution needs to be, and no one has to worry that the change in color won't be seen.

Today, many titrations are done automatically. To do a titration automatically, a pH meter is connected to a machine that adds the base automatically. Some automatic titrators have computers attached to them. The computer regulates the addition of the base. It also stores the pH data from the titration. A graph is plotted out by the system to see when the neutralization takes place. Some systems do not even need humans, except to set up the machine. These titrators can handle many samples. The samples are on a turntable that automatically shifts to the next sample when a titration is finished.

5.5 Where do graphs come from?

One way to analyze data is to use a graph. Graphs give scientists a visual tool that helps them to see patterns in the data they collect. The word "graph" comes from a Greek word *graphein* and means "to write." Scientists use graphs for many purposes.

René Descartes (1596-1650) developed the idea of graphs. Descartes was a French philosopher and scientist. He was so smart that he started college when he was only eight years old. Descartes was sick a lot as a child. Because he was so sick, he spent a lot of time in bed as a child. He did a lot of his thinking while in bed. Descartes believed that science should be as certain as mathematics. He developed the idea of graphs to help support this idea. Graphs allow scientists to actually see scientific data.

Descartes
1596-1650

Today we can plot graphs in a number of ways. We don't even need to buy graph paper anymore. Graph paper can be downloaded from the Internet! Calculators can also plot graphs for us. Graphing calculators automatically plot the graph we want on a window. We don't have to use graph paper if we use a graphing calculator.

The first graphing calculator was invented by a Japanese company in 1985. The lead engineer for the team was **Hideshi Fukaya**. At first, graphing calculators were very expensive. Now, most people can easily afford one.

Graphing calculators for the blind are now available. This calculator makes sounds that the blind person can "see" as a graph. A special printer will print out the graph in Braille.

Another way to make a graph is with a **spreadsheet**. This is a computer program that automatically plots a graph when you put data in the columns. **Dan Bricklin and Bob Frankston** wrote the first spreadsheet program in 1978 while Dan was a college student. Since then, many spreadsheet programs have been on the market.

5.4 Activity

On the last page of this chapter. Complete the time line from 1500 to 1985. On the time line, plot the dates of every discovery, event, and achievement that was discussed in this chapter. If you want, you can look up the exact dates, or you can estimate the dates. Draw a "dot" to represent the time of each discovery, event, and achievement, and write the scientists' names next to the dots.

When you finish the time line, answer the questions on the following page.

1. Explain what you observe on the time line. Describe how both the analyzing of data and the use of graphs help us to understand acid-base reactions.

_____

_____

_____

_____

_____

_____

2. Describe how analyzing data has changed over the course of history.

_____

_____

_____

_____

_____

_____

_____

_____

_____

Time line
graphs

1985

1500

6 Mixtures
History

6.1 Introduction

6.2 Paint mixtures

6.3 Metal mixtures

6.4 Petroleum-hydrocarbon mixtures

6.5 Explosives and fireworks

6.6 Ice cream

6.7 Activity

6.1 Introduction

Mixtures are all around us. They have many useful applications; they are used for products like paint, fuel, and fertilizer. People have used mixtures for centuries, even when the chemistry of the mixtures being used was not understood. Today we know a lot about mixtures and their chemical properties. We know what kinds of molecules mix and what kinds of molecules don't mix. Because we know the chemistry of mixtures, we can create new mixtures for different uses. However, before the chemistry of mixtures was known, people simply tried different mixtures and used the ones that worked.

6.2 Paint mixtures

Painting has been a part of human culture for many centuries. People have painted pictures in many places. Ancient Egyptian walls that were painted over 2000 years ago still have their brilliant color.

Early paints were mixtures of several materials. The ancient Egyptians used mixtures of mineral pigments and possibly beeswax. Oxides of iron, copper, and manganese gave bright colors to the paintings. Sometimes egg, milk, or animals glues were used to hold the pigments.

Many of the early oils used in painting took a very long time to dry. However, the Flemish painter **Jan van Eyck** (1390-1441) used a paint composed of linseed oil and mineral pigments. The linseed oil dried very quickly, thus allowing the painting of different layers.

6.3 Metal mixtures

Many metals are very soft. When a soft metal is bent, it may break with little effort. A soft metal cannot be made into a useful cutting tool, and objects made of soft metals may dent very easily.

An **alloy** is a mixture of two or more elements. At least one of the elements needs to be a metal in order for the mixture to be called an alloy. An alloy is much stronger than a metal by itself. Often two or more metals are mixed to form alloys.

Bronze is an alloy that is a mixture of copper and another metal, such as tin or aluminum. The earliest bronze materials date back to 4000 B.C. and were found in Iran and Iraq. Bronze is much harder than copper. This alloy was widely used for weapons, armor, and tools. One advantage to bronze is that it does not rust.

Steel later became widely used when copper supplies began to run low. Steel is a mixture of iron and small amounts of carbon and metals, such as manganese or tungsten. The use of steel became popular during the Industrial Revolution beginning in the late 1700s.

Silver is a precious metal widely used to make jewelry and eating utensils. Pure silver is very soft, so it cannot be used by itself. Alloys of silver, with small amounts of copper or platinum, make for much sturdier and more useful materials. The use of silver dates back at least to the early Sumerian period, starting around 2900 B.C.

6.4 Petroleum-hydrocarbon mixtures

Petroleum is a mixture of compounds that contain mainly carbon and hydrogen. Petroleum mixtures were used in ancient times. Asphalt made from petroleum was used over 4000 years ago to make the walls of Babylon. Writings from ancient Persian civilizations show that petroleum products were used by the rich people for lighting and for medicines.

It appears that the first oil wells were drilled in the fourth century A.D. by the Chinese. In the eighth century A.D., the people in Baghdad used petroleum from nearby fields to produce tar for paving streets. In the 13th century A.D., **Marco Polo** (1254-1324) wrote about great oil fields as he traveled across Russia.

Petroleum was found in North America in 1595 by **Sir Walter Raleigh** (1554-1618). Commercial development of petroleum products in North America began with the first oil field being drilled in Ontario, Canada, in 1858. **Edwin Drake** (1819-1880) drilled the first U.S. oil well in Pennsylvania in 1859.

The initial commercial products produced by these oil fields were kerosene and oil (for use in oil lamps). The gasoline fraction was thrown away because there was no use for it. When the automobile was invented, gasoline became a very important product.

6.5 Explosives and fireworks

Gunpowder was first discovered by the Chinese around 600-900 A.D. Early alchemists kept records of materials that should not be mixed together. But they still experimented with these mixtures to learn more about their behaviors.

One mixture contained sulfur and saltpeter (potassium nitrate). One record describes how this mixture created a flame that burned the hands of the alchemist. The fire also burned down the shack where the experiments were being performed!

Further experiments showed that the amount of potassium nitrate used in a mixture was a very important factor for obtaining a good explosion. Best results were obtained with a mixture of 75% potassium nitrate, 5% carbon, and 10% sulfur. This mixture is still used today for most gunpowder.

One enjoyable application of mixtures is fireworks. We set off fireworks on the Fourth of July to celebrate Independence Day. Many people also use fireworks to usher in the New Year.

The original firecracker was developed by the Chinese around the same time that gunpowder was developed. The firecracker was first used to scare enemy troops and horses. Gunpowder was later developed into a weapon to harm and kill other people. Rockets, made with gunpowder and bamboo tubes, were used to shoot at enemy soldiers.

Marco Polo brought firecrackers to Europe in 1292. Between 1400 and 1550, the Italians became very interested in firecrackers. They added different metals to produce gold and silver sparks when the fireworks were set off. They also developed rockets that could shoot the fireworks up into the sky.

Modern fireworks can produce a lot of different colors. The addition of strontium to fireworks gives a red color, and the addition of copper will produce a blue color. A green color will be seen when barium is added, and sodium gives fireworks a yellow color. The addition of magnesium and aluminum will produce an even brighter light when the fireworks explode.

6.6 Ice cream

A favorite mixture is ice cream. This delicious mixture was introduced to Europe by Marco Polo in the late 1200s. Modern-day ice cream consists of dairy products, eggs, sugar, stabilizers, air, and flavorings. Thus ice cream is a mixture of solids, liquids, and gases.

Ice cream can also have nuts and fruit and candy in it. It can be a mixture of almost any food item you choose. There is even a Japanese ice cream that contains garlic!

6.7 Activity

Using the dates in this chapter, place beginning and ending dates on the mixtures time line that is on the last page of this chapter. When choosing these beginning and ending dates, be sure to consider the histories of all of the mixtures discussed in this chapter: paint, metals, petroleum, explosives, and (don't forget) ice cream.

On the time line, plot the dates of every discovery, scientist, explorer, event, and painter discussed in this chapter. If you want, you can look up the exact dates, or you can simply estimate them. Using a different color for each mixture, draw a "dot" to represent the time of each event, and then write the name and date next to each dot. (For example, next to one dot, you may write, Bronze 4000 B.C. Next to another dot, you may write, Sir Walter Raleigh 1595.)

When you finish the time line, answer the following questions.

1. Explain what you observe on the timeline. Do you need more information to understand the history of these mixtures? Why or why not?

_____

_____

_____

_____

_____

2. Do you think that understanding the chemistry of mixtures has helped the development of mixtures? Why or why not?

_____

_____

_____

_____

_____

Time line
mixtures

7 Separation History

7.1 Introduction

The ability to separate materials has many useful applications. We can remove impurities from well water in order to obtain good drinking water. We can separate gold and silver from ore, and we can then use the purified metals to make jewelry or silverware. Scientists can also separate useful chemicals from mixtures in plants, and they can use those chemicals to make drugs. Being able to separate mixtures is very valuable for lots of different reasons.

Many useful separation techniques have been discovered by researchers in laboratories. However, many of the ways we separate mixtures were discovered by accident. People just tried different techniques and used what seemed to work best.

7.2 Filtration

A good water supply has always been important for the growth of civilization. Early villages and towns were often located near a water supply. But often, the water was dirty and tasted bad. So people came up with many ways to clean the water supply. One way to clean water is to use some kind of filter. The filter can be anything, such as paper, sand, charcoal, or even gel.

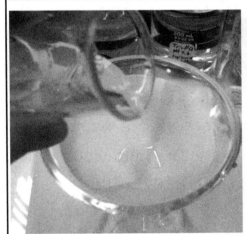 

A water-precipitate mixture is poured through a filter paper.

The precipitate is separated from the water and stays on the filter paper. The clean water is collected below.

Early writings from India (around 2000 B.C.) describe ways that people cleaned water. The water was boiled, or hot metal rods were placed in it. Then people poured the water through sand or charcoal filters to filter the water. These methods removed impurities and gave a better taste.

The Greek physician **Hippocrates** (*circa* 460 - *circa* 377 B.C.) said that pure water was needed for good health. He knew that the water in the Greek aqueducts was not healthy. Hippocrates taught people to clean water by boiling it and then filtering it through cloth.

For centuries, little change took place in water purification. Then during the Renaissance (fourteenth through seventeenth centuries A.D.), people began to study water purity again. The invention of the microscope allowed scientists to see the germs in the water.

The French scientist **Philippe de la Hire** (1640-1718) said that all households should clean their water with sand filters. The citizens of Paisley, Scotland, put in the first city-wide sand filter system in 1804.

Today we still have problems getting completely pure water. Whole house filter systems are used in many places. These devices remove chlorine and other chemicals from the water before the water comes into the house.

7.3 Evaporation

Common table salt (sodium chloride) has played an important role in history for centuries. Salt has been used to flavor food since the beginning of time. Salt also helps preserve food. Some people believe that salt has health benefits. Salt is also important in some religious rites.

The Romans had a well-developed system of transportation for salt. Salt was so important that it was used to pay the Roman soldiers. Wars have been fought over the control of salt supplies. Throughout history, many governments have placed special taxes on salt. These taxes were not popular with the people.

Salt and salt making have been important for cultures everywhere. Egyptian art from the 1400s B.C. shows salt-making methods, and a French drawing from the fifteenth century A.D. shows salt production.

The making of salt by evaporation has a long history. In many places, the heat of the sun dried salty sea water, thereby producing salt that people could use. This process made large amounts of salt. Heating sea water makes the evaporation go faster. Chefs use heat to speed the evaporation process when they make tasty sauces.

7.4 Chromatography

Scientists need to separate materials so that they can study them better. Research on a mixture can give confusing results. If a scientist cannot separate a mixture, he or she may not know which chemical is causing the real effect. Many scientists have worked hard to find good ways to separate materials.

The Russian botanist **Mikhail Tswett** (1872-1919) first developed the science of chromatography. He wanted to study plant pigments. Using chromatography to separate plant pigments, Tswett discovered six different chlorophyll compounds in plants.

Tswett used glass columns filled with chalk to separate the many different colors of materials in plants. This is called **column chromatography.** He was very successful with this technique. Many other scientists later used these same ideas to make great discoveries. The term "chromatography" has come to mean any form of separating materials on a stationary medium.

Tswett's work was published in Russian, so not many people knew about it. He was not a very nice man, and he attacked other scientists and their research. Tswett made a lot of enemies.

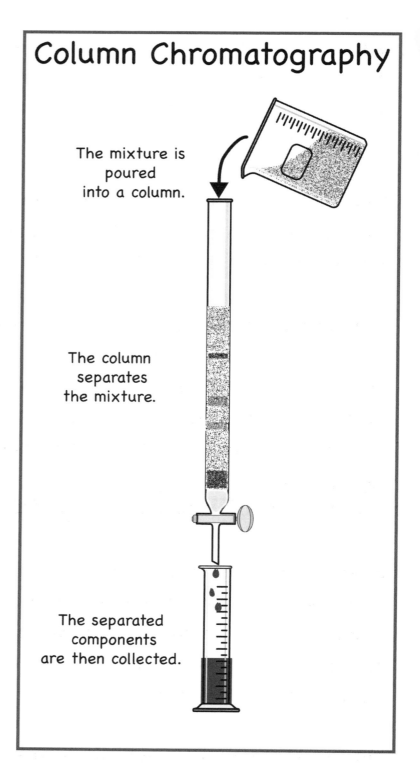

Column Chromatography

The mixture is poured into a column.

The column separates the mixture.

The separated components are then collected.

Other scientists used Tswett's research ideas. **Richard Willstätter** (1872–1942) won the Nobel Prize in Chemistry in 1915 for research that was very similar to that of Tswett.

Another type of chromatography is called **ion-exchange chromatography**. The earliest form of separation using ion-exchange chromatography was developed in the middle 1800s. Two British scientists combined clays and ionic solutions for the purpose of removing certain chemicals. It worked; certain ions attached to the clays and were taken out of the solutions.

This technique was very useful during World War II. The Manhattan Project developed the atomic bomb with the help of ion-exchange chromatography. Ion-exchange chromatography purified different radioactive elements that were needed to create the atomic bomb.

Today we use ion-exchange materials for many purposes. Water softeners in homes employ this technique. Ions such as iron cause "hard water," and by using ion-exchange chromatography, these ions can be removed quickly.

Archer John Porter Martin (1910-2002) continued research on chromatography. He and others developed paper chromatography in 1944. In paper chromatography, the sample solution is applied to a strip of paper. As the liquid moves up the paper, the materials separate.

Paper chromatography helped people learn about the structure of insulin. This hormone is needed to keep blood sugar at the right level. **Frederick Sanger** (1918-present) used both paper and column chromatography to study the structure of insulin. He won the Nobel Prize in Chemistry in 1958 for his research.

7.5 Activity

Using the dates in this chapter, place beginning and ending dates on the time line that is on the last page of this chapter. When choosing these beginning and ending dates, be sure to consider the histories of all of the separation techniques that were discussed in this chapter: filtration, evaporation, and chromatography.

On the time line, plot the dates of every discovery, scientist, and event discussed in this chapter. If you want, you can look up the exact dates, or you can simply estimate them. Using a different color for each separation technique, draw a "dot" to represent the time of each discovery or event, and then write the name and date next to each dot. (For example, next to one dot, you may write the following: Filtering using sand, India, 2000 B.C. Next to another dot, you may write, Archer Martin 1944.)

When you finish the time line, answer the following questions.

1. How does the history of filtration compare to the history of chromatography?

_____

_____

_____

_____

_____

_____

_____

_____

_____

2. What kinds of things can we separate today that we could not
   separate before the invention of chromatography?

_____

_____

_____

_____

_____

3. What advantages or disadvantages do you think Mikhail Tswett
   made for himself by attacking other scientists and their research?

_____

_____

_____

_____

Time line

history of filtration, evaporation, and chromatography

8 Energy & Life History

8.1 Introduction

We have all seen the ads on TV or in the grocery store or on roadside billboards. "Get more energy by drinking product X." "Buy energy bar Y for a better bike ride." "Eat cereal Z for more energy at work." What are these "energy foods," and how do they work?

The body uses two kinds of energy. We use "quick energy" when we need it right away. "Stored energy" is something we have ready to use later when we are not eating.

Quick energy consists of carbohydrates in the bloodstream. Stored energy can be either carbohydrates or fats. We can also sometimes use proteins for energy.

8.2 Basic carbohydrates

Hermann Emil Fischer (1852-1919) was the primary researcher who figured out the structure of the glucose molecule. His research revealed the structure of many carbohydrates. Fischer won the Nobel Prize in Chemistry in 1902 for his studies.

Table sugar (sucrose) has many uses. **Louis Pasteur** (1822-1895) carried out research to show that sugar (sucrose) could be converted to ethanol by yeast. His research was very important for the French beer and wine industries of his day.

Pasteur was one of the most brilliant scientists of all time. He made useful discoveries in many different fields of science. His work made very important contributions to our knowledge of diseases and the prevention of them.

8.3 How the body uses glucose

The major "quick energy" molecule for the body is **glucose**. This compound helps form a chemical called ATP that stores chemical energy. When a piece of the ATP molecule is broken off, chemical energy is released. This energy can be used by the cells to do useful work.

Many scientists helped us learn about the process known as **glycolysis** (glucose splitting). Along with the discoveries by Louis Pasteur, it took many men and women, and over one hundred years, to work out the process.

Eduard Buchner (1860-1917) studied how the cells use glucose. Buchner was a German scientist who studied both chemistry and botany. He showed that living cells were not needed to change glucose, but that glucose could be changed in a laboratory.

Buchner broke up cells and pressed the juice out of them, and he found that this "juice" could convert glucose to other compounds. He won the Nobel Prize in Chemistry in 1907 for his research.

However, nobody knew what was in the juice. What compound was making the glucose change? In 1905, two scientists found the answer.

Arthur Harden (1865-1940) and **William Young** (1878-1942) discovered that an enzyme (which is a protein) and a set of small molecules make the reactions happen.

The juice causes a complicated set of reactions to occur. These reactions change glucose to a compound called pyruvic acid. There are many steps to this reaction series.

In biochemistry, a set of reactions like these are sometimes referred to as a pathway. The glycolysis pathway was developed by two scientists, **Gustav Embden** (1874-1933) and **Otto Meyerhof** (1884-1951). Gustav Embden was a German chemist. He worked on muscle contraction studies. Otto Meyerhof, also German, was a physician and a biochemist. Meyerhof shared the Nobel Prize in Medicine in 1922 for his research on lactic acid formation. Because of their work, we now call the set of reactions in the juice the **Embden-Meyerhof Pathway**.

8.4 The Krebs cycle

The formation of pyruvic acid is just the start of energy production. In order to make ATP, two more major processes are needed. These processes are the Krebs cycle and the method for actually making ATP.

Hans Krebs (1900-1981) was born in Germany and later moved to England. He was trained as both a physician and a biochemist. His major work was figuring out the cycle that was named after him. In 1953, Krebs won the Nobel Prize in Physiology or Medicine for his research, and in 1958, he became a knight.

The Krebs cycle is a series of reactions that goes in a circle. At several places in the cycle, NADH is made. This compound is then used in the third phase of ATP manufacture. Other useful compounds are also made in this cycle.

8.5 Making ATP in the cell

Now let's put it all together. First glucose is changed to pyruvic acid in the cell. The pyruvic acid goes into the mitochondria (little particles inside the cell). Inside the mitochondria, the pyruvic acid helps make NADH. Then, in a very complicated process, the NADH is used to make ATP.

For a long time, we did not understand how the NADH helps to form ATP. There were several theories and some heated arguments about what actually happens. But eventually, the details were worked out.

Peter Mitchell (1920-1992) discovered that hydrogen ions aid the process. They move inside the mitochondria to help form ATP. Mitchell won the Nobel Prize in Chemistry in 1978 for his studies.

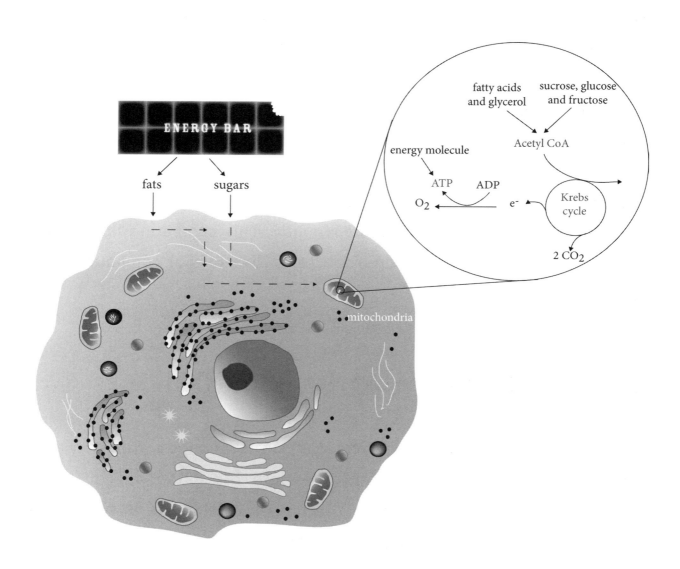

The final piece of the puzzle was put together by **Paul Boyer** (1918-present). He discovered the enzyme that puts the ATP molecule together. For this work, Boyer was awarded the Nobel Prize in Chemistry in 1997.

And this is how your body uses the "energy food" that you buy from the grocery store. It is a long and complicated process that involves many different reactions and pathways. No wonder it took so long to figure out!

8.6 Activity

Using the dates in this chapter, place beginning and ending dates on the time line that is on the last page of this chapter. When choosing these beginning and ending dates, be sure to consider the entire history of glycolysis. Your time line should begin with the discovery of glucose, and it should end with the Krebs cycle (and how ATP is produced).

On the time line, plot the dates of every scientist in this chapter who was involved in the history of glycolysis. If you want, you can look up the exact dates, or you can simply estimate them. Draw a "dot" to represent the time of each discovery, and write the appropriate name and date next to each dot.

When you finish the time line, answer the questions on the following page.

1. Explain what you observe on the time line regarding the dates of experimental discoveries. Are these dates clustered, or are they spread over many years?

_____

_____

_____

_____

_____

2. Compare this time line to other timelines in this workbook. Describe any differences or similarities you observe.

_____

_____

_____

_____

3. Do you think that knowing about glycolysis has changed the way we eat? Why or why not?

_____

_____

_____

_____

Time line
the history of glycolysis

9 Polymers
History

9.1 Introduction

European explorers came across the Mayans of Central America in the 1500s. They saw the Mayan children playing with balls. These balls were made of materials from a local tree. This is the first known report of rubber.

The Mayans got their rubber by using a substance called latex. Latex is the milky sap that comes from rubber trees.

Use of rubber by the Mayans dates back to about 1500 B.C. They used rubber for many purposes. They used it to make balls and hollow dolls. The Mayans also used rubber to tie axe heads to handles for chopping and for use in battle.

9.2 Rubber comes to Europe

Spanish and Portuguese writers described rubber in the 1500s. However, there was little interest in this new material. Two French scientists, **Charles Marie de la Condamine** (1701-1774) and **Francois Fresneau** (1703-1770), studied rubber. Their reports to the French Academy of Sciences in the mid-1700s brought rubber to the attention of many people.

At that time, rubber was used for only a few purposes. Sheets of rubber from Brazil were used to make rubber bands and waterproofing materials. **Joseph Priestly** (1733-1804) invented the rubber eraser in 1770.

In the 1800s, the use of rubber began to grow. Rubber factories were set up in Paris in 1803 and in England in 1820. **Charles Goodyear** (1800-1860) invented the vulcanization technique in 1839. This process involved adding sulfur to the liquid rubber to make it harder.

During World Wars I and II, ways of making synthetic rubber were developed. These methods were needed because we did not have good natural rubber sources. Today about 60% of the rubber we use is synthetic.

9.3 Understanding polymer structure

It took a long time for scientists to understand the structure of polymers. We had some idea of how natural polymers, like starch, were put together. But chemists did not get very interested in other polymers.

The English chemist **Charles Greville Williams** (1829-1910) distilled natural rubber. He found a compound that he called isoprene in 1860. This molecule had only five carbon atoms and eight hydrogen atoms—a very small compound.

Twenty years later, the French scientist **Gustave Bouchardat** (1842-1919) used isoprene to make a compound like rubber. His work strongly suggested that natural rubber was composed of isoprene. But he did not know how the isoprene molecules were put together to make rubber.

The German scientist **Carl Harries** (1866-1923) did research on this question. He started writing about his ideas in 1904. Harries believed that the isoprene chains were somehow able to link end-to-end to make rubber.

The German organic chemist **Hermann Staudinger** (1881-1965) began work in 1920 to convince others of the existence of long-chain molecules. Many other scientists started to accept the idea. They used many techniques to study the properties of these molecules. They learned the structures of these molecules, and they discovered how new long-chain molecules could be made.

9.4 Polymer products

Many natural polymers exist. We have already talked about the carbohydrate polymers: starch and cellulose. More natural polymers will be discussed in the next chapter.

Cellulose nitrate was one early combination of a natural polymer and other chemicals. The French plant chemist **Henri Braconnot** (1780-1855) reacted nitric acid with starch (wood chips). He developed the first guncotton, but it was not very good.

The first useful guncotton was discovered accidentally. **Christian Friedrich Schönbien** (1799-1868) was a German-Swiss chemist. One day, he spilled some nitric acid in his kitchen. He wiped up the spill with a cotton towel. When the towel dried, it exploded. This combination of materials was used for a while as an explosive. But it was too dangerous to manufacture. So other explosives were later developed.

rubber

polyethylene

polyester

nylon

Many synthetic polymers were created in the twentieth century. These polymers were not found in nature. They were created in the lab, and they had many uses.

The first synthetic polymer was made in 1907. **Leo Baekeland** (1863-1944) developed this material. He named it Bakelite. This polymer was very hard, and it resisted heat.

Baekeland wanted to find a way to replace shellac. Shellac, a natural polymer, was made from the shells of oriental lac beetles (get it? – shel + lac = shellac). After he developed Bakelite, he sold the rights to manufacture the product to Union Carbide. Then he retired and went sailing on his yacht.

The synthetic polymer PVC (polyvinyl chloride) is used for plumbing, pipes, plastic bottles, and a number of other products. PVC was discovered separately by **Henri Victor Regnault** (1810-1878) in 1835 and also by **Eugen Baumann** (1846–1896) in 1872. However, this product could not be used commercially, and it existed mainly as tiny flakes. In 1926, **Waldo Semon** (1898–1999) and the Goodrich Corporation developed a method to turn the flakes into a solid plastic by blending them with other chemicals. Once blended with other chemicals, the PVC plastic was more flexible, and it could then be used for commercial products.

Nylon and Kevlar are two important synthetic polymers. Nylon was a great invention for the clothing industry. This material was developed by the DuPont Company in 1938. Nylon has a more complicated structure than some other polymers, and it is now widely used for making clothing.

Kevlar was developed in 1965 by two researchers who worked for DuPont: **Stephanie Kwolek** and **Roberto Berendt**. This material's strength and its ability to withstand high temperatures make it very useful for a variety of products, including bulletproof vests and fireproofing materials.

9.5 Activity

Using the dates in this chapter, place beginning and ending dates on the polymers time line that is on the last page of this chapter. When choosing these beginning and ending dates, be sure to consider the histories of all of the polymers (both natural and synthetic) that were discussed in this chapter.

On the time line, plot the dates of the discoveries that were made by every scientist discussed in this chapter. If you want, you can look up the exact date of each discovery. Or you can estimate the dates by assuming that the discoveries took place sometime in the middle of the scientists' lifetimes. Draw a "dot" to represent the time of each discovery, and write the scientists' names next to the dots. Also, note whether they worked on natural or synthetic polymers.

When you finish the time line, answer the questions on the following page.

1. Describe what you notice regarding the time span between the discovery of and the use of natural and synthetic polymers.

_____

_____

_____

_____

_____

_____

_____

2. Notice how difficult it is to pinpoint the date of a discovery. Was PVC "discovered" by Regnault and Baumann, or by Semon? Explain.

_____

_____

_____

_____

_____

_____

_____

Time line
polymers

10 Proteins & DNA History

10.1 Introduction

Alfred Nobel (1833-1896) invented dynamite and made a lot of money. In his 1895 will, he established awards for work in science and in the humanities. Today, each of these annual awards (a Nobel Prize), is often over one million dollars.

Wouldn't you like to win a Nobel Prize? If you do research in protein or DNA chemistry, you might have a good chance. Many Nobel Prizes have been given for work in these two areas of science.

10.2 What are proteins?

The word "protein" comes from a Greek word *prota*. It means "of primary importance." Proteins were the first very large molecules to be studied chemically in detail.

The idea of proteins was first developed by the Swedish physician and chemist **Jöns Jakob Berzelius** (1779-1848). Soon after Berzelius introduced his ideas, many proteins were analyzed for their chemical content. Scientists found that all proteins seemed to have the same general chemical makeup.

Berzelius and others of his time associated proteins with living systems. But Berzelius did not accept the idea of vitalism. He believed that there was a force that somehow organized tissues in an organism.

Proteins were soon thought to have something to do with animal nutrition. Measurement of protein in urine was found to be medically useful by 1905. But a much more useful role for proteins was soon thereafter discovered.

In 1926, **James Sumner** (1887-1955) discovered how proteins make reactions occur. He studied the reaction in bacteria that converts urea into carbon dioxide and ammonia. Sumner showed that the material responsible for the reaction was a protein. (We call this kind of protein an **enzyme**.) Sumner, at age 17, had to have his left arm amputated below the elbow due to a hunting accident. He was left-handed, and thus, he had to learn to do everything with his right hand. Despite his disability, he went on to share the Nobel Prize in Chemistry in 1946 for his enzyme studies.

Many other enzymes were soon discovered. Now we know of more than fifteen hundred different types of enzymes. Enzymes carry out many complicated reactions in living systems.

10.3 What do proteins look like?

Proteins are composed of amino acids. The amino acids connect to one another to form a long chain. Sometimes two or more protein chains will connect with one another to make the complete protein molecule.

Frederick Sanger (1918-present) studied how the amino acids were connected in the protein insulin. He used chemical reactions to remove each amino acid from the chain. Chromatography allowed him to separate the amino acids so that he could learn the sequence in the insulin protein. Sanger won his first Nobel Prize in Chemistry in 1958 for this work.

Proteins are molecules with three-dimensional shapes. X-rays were used to learn about these shapes. Beams of X-rays were sent through protein crystals, and this technique helped us learn what proteins look like. Two similar proteins were the first ones to be studied with this technique: hemoglobin and myoglobin. **Max Perutz** (1914-2002) studied the protein hemoglobin, which carries oxygen in red blood cells. **John Kendrew** (1917-1997) did research on myoglobin. This molecule carries oxygen in muscles. Perutz and Kendrew shared the 1962 Nobel Prize in Chemistry for their work.

10.4 DNA—the software of a cell

Another large molecule is DNA. This molecule serves as the "library" for cell information. The library has directions for how to build the cell and for what the cell will do.

Early ideas about how we inherit eye color and other traits were developed by **Gregor Mendel** (1822-1884). Mendel was a priest who studied pea plants. He discovered some laws about inheritance that allow us to predict certain things.

Later work on DNA was made possible when the microscope was invented. With a microscope, we could finally see inside the cell, and we could look at chromosomes (the structures that control inheritance). Chromosome studies showed that both proteins and DNA were involved in cell development. But what was really happening? Was it the complex protein (with over twenty amino acid components) or the DNA (with only four components) that was responsible for making new cells?

Alfred Hershey (1908-1997) and his lab assistant, **Martha Chase** (1927-2003), provided the answer to the question. In 1952, they found that DNA grew inside the cell to help make new cells. For this work, Hershey shared the Nobel Prize in Physiology or Medicine in 1969 with two other researchers.

10.5 The structure of DNA

Following the work of Hershey and Chase, we knew something about what DNA did inside the cell, but we did not have any ideas about how the molecule worked. In order to understand the function, we needed to know the structure of DNA.

Many people were working on this problem. The X-ray diffraction technique had been developed for proteins and other molecules, and scientists began using this approach to study DNA.

James Watson (1928-present) and **Francis Crick** (1916-2004) came up with the double helix structure for DNA in 1953. They worked with wire and cardboard models. Their structure had the bases inside the chain. They were awarded the 1962 Nobel Prize in Physiology or Medicine for their research.

One unfortunate part of this history was how Watson and Crick got some of their data. It was given to them by another researcher who took it from **Rosalind Franklin** (1920- 1958). Franklin was a British biochemist and X-ray crystallographer. Franklin had good X-ray diffraction data of DNA, and because her data was so good, Watson and Crick were able to predict the structure. Watson and Crick used her data without asking her permission. Many people feel she should have also received the Nobel Prize for her contribution. Many years later, Watson finally admitted taking the data.

The nucleotide sequence was developed in the 1970s by **Frederick Sanger** (the insulin structure man). He won a second Nobel Prize in Chemistry in 1980 for this work.

10.5 Activity

Using the dates in this chapter, place beginning and ending dates on the protein and DNA time line that is on the last page of this chapter.

On the time line, plot the dates of the discoveries that were made by every scientist discussed in this chapter. If you want, you can look up the exact date of each discovery. Or you can estimate the dates by assuming that the discoveries took place sometime in the middle of the scientists' lifetimes. Draw a "dot" to represent the time of each discovery, and write the scientists' names next to the dots.

When you finish the time line, answer the questions on the following page.

1. There are literally thousands of different proteins in living things. Each one was discovered by one or many scientists. Look at the cluster of dates for discoveries of proteins and DNA, and compare this cluster to the time line in chapter eight. What do you notice?

_____

_____

_____

_____

_____

_____

_____

_____

2. If you combine the dates in chapter eight with those in chapter ten, you can see lots of new discoveries happening in a very short period of time. What historical, technological, and philosophical factors may have contributed to this explosion of research?

_____

_____

_____

_____

_____

3. Explain why you think that Rosalind Franklin was not given credit for her important work. Explain how things may or may not have changed for women in science today.

_____

_____

_____

_____

_____

_____

_____

_____

_____

_____

_____

_____

_____

_____

_____

_____

Time line
protein and DNA

Time line

Time line

Time line